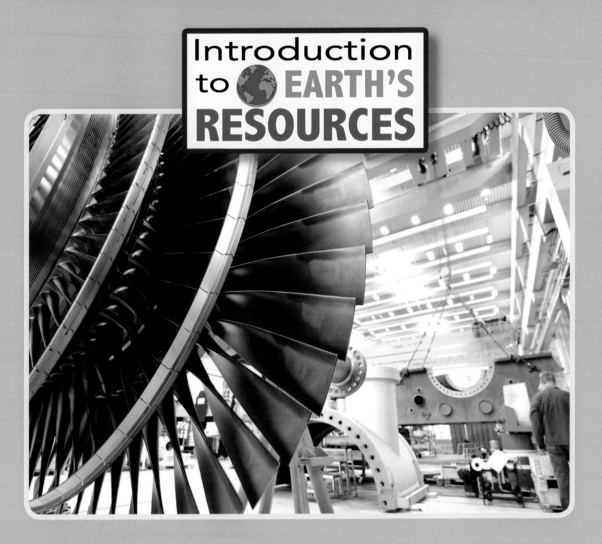

Introduction to EARTH'S RESOURCES

HOW WE USE
FOSSIL FUELS

Nancy Dickmann

Crabtree Publishing Company
www.crabtreebooks.com

Crabtree Publishing Company
www.crabtreebooks.com

Author: Nancy Dickmann
Editorial Director: Kathy Middleton
Editor: Ellen Rodger
Picture Manager: Sophie Mortimer
Design Manager: Keith Davis
Children's Publisher: Anne O'Daly
Proofreader: Debbie Greenberg
**Production coordinator and
  Prepress technician:** Ken Wright
Print coordinator: Katherine Berti

Photographs (t=top, b=bottom, l=left, r=right, c=center)
Front Cover: All images Shutterstock
Interior: iStock: banksPhotos 15, Hal Bergman 19, George Clerk 13, ewg3D 11, Extreme Photographer 27, HellRy 12, industryview 1, 17, IP Galanternik D.U. 14, Kstevecope 23, Olivier Le Moal 28b, mbbirdy 5, Don Mennig 25, Mimadeo 26, small smiles 10, Michael Utech 21, xxmmxx 22, tony Yao 8, Qin Zhao 24; Shutterstock: AdShooter 6, Andrei Minsk 9, Sergey Panychev 29cl, Piscfive 29br, romrodphoto 28t, Frederico Rogagno 20, Stockr 16, VectorMone 7, Wisit Tongma 4.
All facts, statistics, web addresses and URLs in this book were verified as valid and accurate at time of writing. No responsibility for any changes to external websites or references can be accepted by either the author or publisher.

Library and Achives Canada Cataloguing in Publication

Title: How we use fossil fuels / Nancy Dickmann.
Other titles: Fossil fuels
Names: Dickmann, Nancy, author.
Description: Series statement: Introduction to Earth's resources |
  Includes bibliographical references and index.
Identifiers: Canadiana (print) 20200283987 |
  Canadiana (ebook) 20200284029 |
  ISBN 9780778781882 (softcover) |
  ISBN 9780778781820 (hardcover) |
  ISBN 9781427126009 (HTML)
Subjects: LCSH: Fossil fuels—Juvenile literature. |
  LCSH: Fossil fuels—Environmental aspects—Juvenile literature. |
  LCSH: Pollution—Juvenile literature.
Classification: LCC TP318.3 .D54 2020 | DDC j333.8/2—dc23

Library of Congress Cataloging-in-Publication Data

Names: Dickmann, Nancy, author.
Title: How we use fossil fuels / Nancy Dickmann.
Description: New York, NY : Crabtree Publishing Company, 2021. |
  Series: Introduction to earth's resources | Includes index.
Identifiers: LCCN 2020029721 (print) | LCCN 2020029722 (ebook)
  ISBN 9780778781820 (hardcover) |
  ISBN 9780778781882 (paperback) |
  ISBN 9781427126009 (ebook)
Subjects: LCSH: Fossil fuels--Juvenile literature.
Classification: LCC TP318.3 .D535 2021 (print) |
  LCC TP318.3 (ebook) | DDC 553.2--dc23
LC record available at https://lccn.loc.gov/2020029721
LC ebook record available at https://lccn.loc.gov/2020029722

Crabtree Publishing Company
www.crabtreebooks.com   1-800-387-7650
Published in 2021 by Crabtree Publishing Company

Copyright © Brown Bear Books Ltd 2020

Published in Canada
Crabtree Publishing
616 Welland Ave.
St. Catharines, ON
L2M 5V6

Published in the United States
Crabtree Publishing
347 Fifth Ave
Suite 1402-145
New York, NY 10016

Printed in the U.S.A./082020/CG20200710

In Canada: We acknowledge the financial support of the Government of Canada through the Canada Book Fund for our publishing activities.

Contents

What Are Fossil Fuels?

Fossil fuels are always in the news. They are very useful, but their use also causes problems.

Fossil fuels is a term for several different kinds of **fuel**. These fuels are made from plants and animals that died millions of years ago. As they broke down, they formed fuels in the ground.

Workers drill deep below the surface to reach some fossil fuels.

About **80%** of all the **energy** used around the world comes from fossil fuels.

Without fossil fuels to provide energy, our world would look very different.

Types of Fuel

The three main types of fossil fuels are coal, oil, and natural gas. These three fuels are different, but they all contain a lot of energy. Fossil fuels must be burned to release this energy. We can use it to produce **electricity**. We can also use it to power vehicles and machinery.

Most of the world's oil formed between **252** and **66 million years** ago, at the same time as the dinosaurs lived.

How Fossil Fuels Form

It takes millions of years for fossil fuels to form. It happens deep underground.

Coal is made from the remains of **prehistoric** plants. When they died, they were covered over with mud. Over millions of years, they got buried deeper under even more layers. The mud turned into rock and the plants slowly changed into coal.

Coal is often found in layers, just like it was formed.

The first oil well in Texas was dug in 1866. It reached oil **106 feet** (32 m) below the surface.

In 2009 an oil well in the Gulf of Mexico reached a depth of **35,000 feet** (10,700 m).

1 Living things die and sink to the bottom.

2 Layers of soil, sand, or rock build up over them.

3 The layers turn into rock. The plants and animals turn into oil or natural gas.

Many of today's land areas were once covered by oceans.

Same but Different

Oil and natural gas form in a similar way, but they are made from the remains of plants and animals that lived in the ocean. When they died, they settled on the ocean floor. Then they were covered over. For all types of fossil fuels, the heat and **pressure** caused by the layers of rock above them turn the dead material into fuel.

Where in the World?

Some places on Earth are rich in fossil fuels. Others have hardly any.

Most countries have at least some coal buried underground. Russia, Australia, and the United States have a lot. These regions were once covered with lush prehistoric swamps. When the plants in these swamps died, over time they were turned into coal.

China produces more coal than any other country.

In 2018 China produced **3.8 billion tons** (3.4 billion metric tons) of coal. This is **4 1/2** times as much as the second-place country.

About **12%** of the world's oil comes from Saudi Arabia. They produce about **12 million barrels** each day.

These are the world's top ten oil-producing countries.

Rare and Precious

Fewer countries have oil and natural gas. Saudi Arabia has a lot of both, and so do Russia and the United States. Norway is also a major producer of natural gas. Many supplies of oil and gas are buried under the oceans. People build **offshore rigs** on the water to drill down to these fossil fuels.

1. United States
2. Saudi Arabia
3. Russia
4. Canada
5. China
6. Iraq
7. Iran
8. United Arab Emirates
9. Brazil
10. Kuwait

Coal

Coal is a solid fuel that looks like a shiny black rock. It is mainly made up of carbon.

People have burned coal for thousands of years. When it burns, it releases a lot of energy—more than wood does. In the 1700s, burning coal heated the water to power **steam** engines. In the 1800s, people also began to use coal to heat their homes.

Many people around the world still burn coal to heat their homes.

Wood contains two-thirds as much energy as coal. Oil has nearly two times as much.

Digging out coal can damage the landscape.

Coal Today

Steam engines were eventually replaced by **gasoline** engines. Today, most coal is used for generating electricity. Some coal deposits are near the surface. Miners can use heavy machinery to scrape off the layers above, then dig the coal out. Deposits that are deeper underground have to be reached by digging tunnels.

The shaft in a coal mine can go down more than **1,000 feet** (305 m).

Oil

Oil is a thick black liquid. It is found in pockets called reservoirs, deep underground.

In a few places, oil naturally seeps up onto the surface. Ancient people used this oil. People didn't start drilling into the ground for oil until the 1800s. It was useful as a fuel. It was also a good lubricant, which is a slippery substance used to keep machine parts moving smoothly.

Oil is shipped around the world in huge boats called tankers.

Around the world, we use about 100 million barrels of oil each day.

About 43% of a typical barrel of oil is made into gasoline. About 9% is turned into jet fuel.

Modern oil refineries are huge and full of complicated parts.

One Fuel, Many Uses

People often use the term **crude oil** and petroleum. Crude oil refers to the oil as it is found underground. Petroleum includes any of the products that are made from it. In a factory called a refinery, different parts of the crude oil are separated out. They are turned into gasoline, jet fuel, and other products.

13

Natural Gas

Coal is a solid and oil is a liquid. What is natural gas? A gas, of course!

Natural gas is often found in the same underground pockets as crude oil. The oil sits below, and the gas rises to the top. Some oil wells produce both oil and natural gas. There are also some pockets that contain only gas and no oil.

The burners on a stove are sometimes fueled by natural gas.

In the United States, about **17%** of natural gas is used in homes. **28%** is used in factories.

Anyone who works on gas pipes must check carefully for leaks.

Making Gas Safe

Natural gas will catch fire or cause an explosion if it is near a flame or a spark. For this reason, gas leaks can be very dangerous. Natural gas has no color or odor, so it can be hard to detect. A smelly chemical is often added to gas supplies. This makes it easier to tell if there is a gas leak.

In 2018 world gas use jumped by **4.6%**, compared to the previous year.

Lighting the World

One of the main uses of fossil fuels is for producing electricity.

We use electricity every day. It powers lights, fridges, computers, and even cars. There are many different ways to generate, or create, electricity. Some methods use sunlight. Others use wind or flowing water. But most of our electricity comes from burning fossil fuels.

Most electricity is produced in factories called power plants.

About **65%** of the world's electricity is produced by burning fossil fuels.

In most power plants, only about **40%** of the energy in coal is converted to electricity.

Electricity is sent through wires to wherever it is needed.

Steam turbines look like huge fans with many blades.

How It Works

In a power plant, fuel is burned to release its energy. The energy heats up water until it turns into steam. The steam turns the blades of a giant **turbine**, and the spinning of the turbine is turned into electricity. Most of these plants burn coal or natural gas. Very few use oil as a fuel.

Fueling Transportation

The energy in fossil fuels has been powering transportation for many years.

Burning coal provided the steam to run early train engines. In the late 1800s, engineers developed engines that could run on gasoline. Since then, the world has come to depend on this fuel, which is made from crude oil.

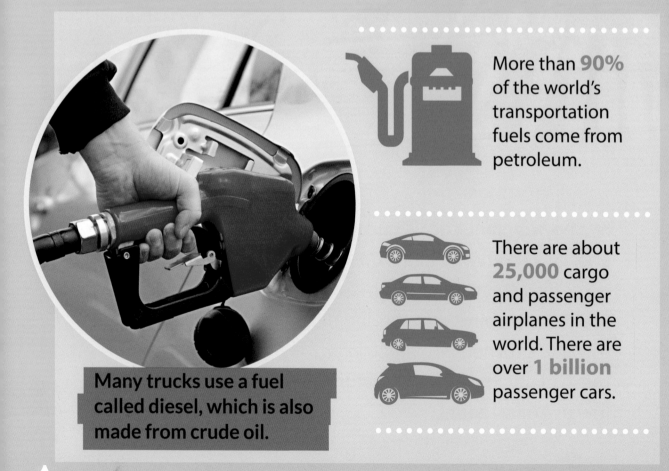

More than **90%** of the world's transportation fuels come from petroleum.

There are about **25,000** cargo and passenger airplanes in the world. There are over **1 billion** passenger cars.

Many trucks use a fuel called diesel, which is also made from crude oil.

Ships use thick fuels that are made from crude oil.

Planes and Trains

Airplanes rely on a fuel that is made from crude oil. Jet fuel won't freeze at the low temperatures high up in the air like gasoline would. Many train engines use diesel fuel. Some trains are electric, but the electricity that powers them often comes from burning fossil fuels. So does the electricity used to power electric cars.

Are Fossil Fuels Renewable?

We use a lot of fossil fuels for electricity and transportation. Will we ever run out?

Some types of energy are **renewable**. This means that they are based on something that can be replaced. For example, wood is renewable. If we burn logs on a fire, the wood gets used up. But we can plant a new tree to replace it.

Solar power is energy from the Sun. It is renewable energy because the Sun will not run out.

About **28%** of the world's electricity comes from renewable sources such as wind, solar, and water.

Renewable energy makes up about **15%** of the world's total energy use.

Every bit of fossil fuels that we burn means there is less left.

Running Out

Fossil fuels are not renewable. Once we use up the current supply of coal, oil, and natural gas, there will be no more. It would take millions of years for more to form deep underground. Scientists think we will run out of oil and natural gas before we have used up all the world's coal.

Polluting the Air

Fossil fuels are not renewable. But that's not the only problem with these fuels.

When we burn fossil fuels, they release gases into the air. Some of these gases are harmful to people and wildlife. Breathing them in can cause health problems. Burning fossil fuels also causes **smog** and **acid rain**.

Acid rain has chemicals that can damage trees and harm wildlife.

One study linked **pollution** from coal plants in the United States to **13,200 deaths** in a single year.

Waste from coal mines can get into streams and rivers.

Harming the Planet

Getting fossil fuels out of the ground can also cause pollution. Oil can leak out, which pollutes water and is dangerous to wildlife. Some chemicals used in mining and drilling can be dangerous if they get into the environment. Digging and drilling for fossil fuels can destroy important **habitats**, harming animals.

An explosion on an oil rig in 2010 released 4.9 million barrels of oil into the ocean.

Warming the Planet

Carbon dioxide is one of the gases released when fossil fuels burn. Too much is heating up the planet.

Carbon dioxide (or CO_2) is a gas found naturally in the air around us. Humans and animals breathe it out, and plants take it in. In fact, plants need it to survive! But burning fossil fuels releases so much CO_2 that it is building up in Earth's **atmosphere**.

Even huge forests can't take in all the carbon dioxide that humans are producing.

Since 1950 the amount of CO_2 in the atmosphere has jumped by about **one-third**.

Higher temperatures melt sea ice. This causes ocean levels to rise.

Global Warming

The extra CO_2 rises into the atmosphere and forms an invisible layer. It acts like the glass in a greenhouse, trapping the Sun's heat. Instead of bouncing off Earth and back into space, the trapped heat is slowly warming Earth. It is causing the entire planet's **climate** to change.

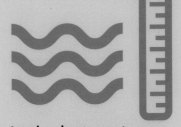

In the last century, Earth's average temperature increased by about 1.62 °F (0.9 °C). Ocean levels rose by about 8 inches (20 cm).

What's the Alternative?

We need to switch from using fossil fuels. What can we use to replace them?

We are making progress on replacing fossil fuels to make electricity. Some methods of producing electricity are renewable and do not pollute, such as wind power and solar power. Hydroelectric power, which turns the energy of flowing water into electricity, is also widely used.

Wind farms can produce enough energy to power a small city.

Scientists estimate that it would take **4,000** large wind turbines to power New York City.

New Ways to Travel

However, replacing fossil fuels for transportation will be harder. Most vehicles still use fossil fuels, although electric cars are becoming more common. There are already electric trains. Engineers are designing electric trucks. But it will be years before we have the technology to replace passenger planes with electric ones.

About **2%** of the new cars sold in the United States are electric. In Norway, it is **39%**.

Some of the electricity used to charge electric cars still comes from burning fossil fuels.

27

What Can I Do?

Fossil fuels are useful, but harmful to the planet. Here are some tips on how you can use less.

- Cut down on car use, especially for short journeys. Could you walk, ride a bicycle, or take a bus instead?

- Find out where your electricity comes from. You may be able to choose a renewable source instead.

- Use less electricity by turning off lights and gadgets when not in use.

- Don't turn the heat on too high in winter. Put on a sweater instead!

- Plastics are made from fossil fuels. Try to avoid buying disposable plastic items such as bottled water.

Quiz

How much have you learned about fossil fuels? It's time to test your knowledge!

1. What are the three main types of fossil fuels?

a. wind power, solar power, and hydroelectric

b. coal, oil, and natural gas

c. dinosaurs, mammoths, and cavemen

2. Where are fossil fuels formed?

a. deep underground

b. in a power plant

c. in volcanoes

3. What is gasoline made from?

a. natural gas

b. crude oil

c. animal waste

4. What happens to extra CO_2 in the atmosphere?

a. birds breathe it in

b. it turns into puffy white clouds

c. it forms an invisible layer that traps the Sun's heat

5. Why do we add smelly chemicals to the gas supply?

a. to make it smell like chocolate

b. to make it renewable

c. to make it easier to sniff out gas leaks

Answers on page 32.

Glossary

acid rain Rain that has mixed with polluting gases to become harmful

atmosphere The blanket of gases that surround Earth

carbon dioxide A gas found naturally in the atmosphere, which is also produced when we burn fossil fuels

climate The average weather patterns in an area over a long period of time

crude oil A fossil fuel that comes out of the ground and can be turned into many different fuels

electricity The flow of current that we use to run gadgets, motors, lights, and more

energy The ability to do work

fuel A substance that is burned to release energy

gasoline A fuel that is made from crude oil and is often burned in vehicle engines

habitats The natural homes of animals or plants

offshore rigs Platforms in the ocean that hold equipment for drilling for oil

pollution The dirtying of water, air, or other environments by harmful substances

prehistoric The time before history was written down

pressure A pushing force

renewable Able to be replaced rather than being used up entirely

smog Fog or mist that is mixed with smoke or pollution

steam Water in the form of a gas

turbine A device with blades that spin when a gas or liquid flows past it

Find out More

Books

Chambers, Catherine. *How Harmful Are Fossil Fuels?* (Earth Debates). Heinemann Library, 2015.

Dickmann, Nancy. *Burning Out: Energy from Fossil Fuels* (Next Generation Energy). Crabtree Publishing, 2016.

Grady, Colin. *Fossil Fuels* (Saving the Planet Through Green Energy). Enslow Publishing, 2017.

Wang, Andrea. *How Can We Reduce Fossil Fuel Pollution?* (Searchlight Books). Lerner Publications, 2016.

Websites

www.bbc.co.uk/bitesize/topics/zshp34j/articles/zntxgwx
Visit this website to find out about fossil fuels and renewable energy.

climate.nasa.gov/evidence/
This NASA website has facts and information about climate change.

www.eia.gov/energyexplained/
Go to this website for information about renewable and nonrenewable sources of energy.

Index

Quiz answers
1. b; 2. a; 3. b; 4. c; 5. c